INTRODUCTION

This book is intended for children between the ages of 6 and 9, and it is said to be very beneficial for children in school settings to dramatically enhance their arithmetic ability by putting them to the test establishing a foundation of skills and knowledge for addressing mathematical problems. These exams will assess students' growth and accomplishment as well as the efficiency of the resources and pedagogical techniques they employed in their respective schools. Ascending and descending order, shapes, easy addition, division, subtraction, and multiplication exercises, as well as pertinent examples and tasks to test young readers, are all included in this book.

MODULE 1

Example 1:

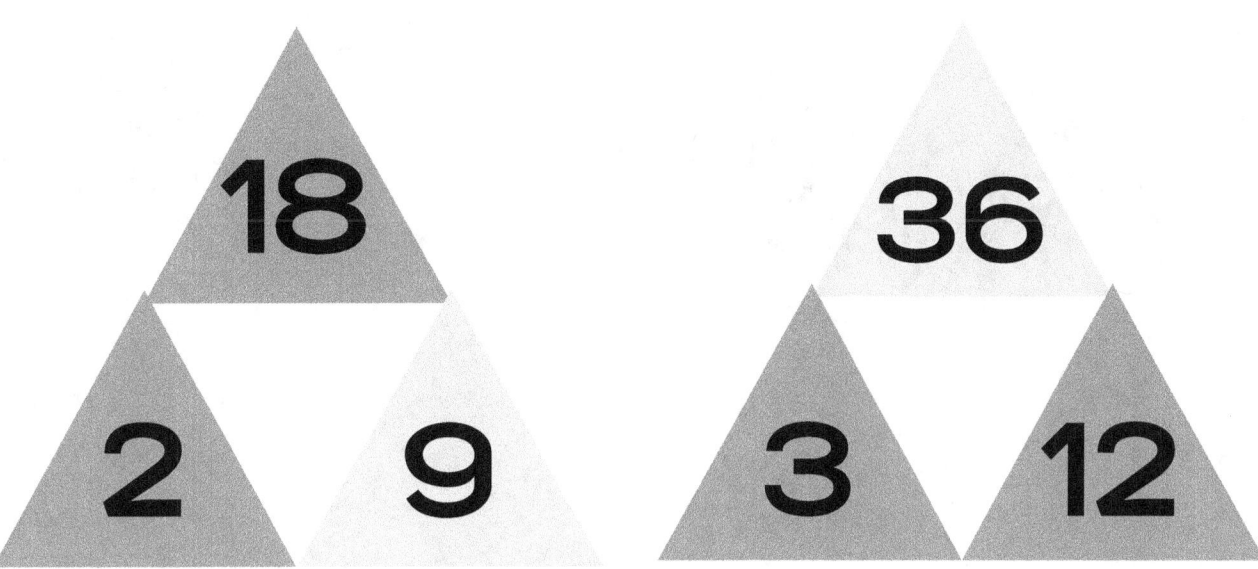

Exercise 1:

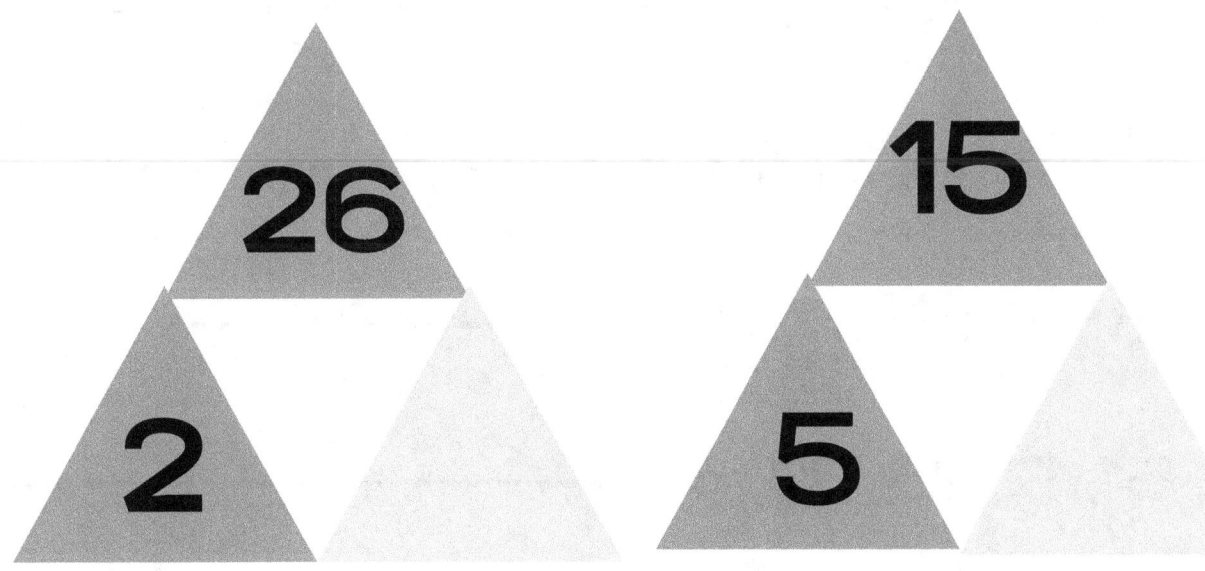

MODULE 2

Exercise 2 :
ARRANGE THESE NUMBERS IN ORDER FROM DESCENDING TO ASCENDING FORM :

1. 26, 2, 14, 17, 9 ------------------------------

2. 4, 3, 15, 6, 176 ------------------------------

3. 14, 33, 7, 56, 8 ------------------------------

4. 5, 95, 152, 2, 34 ------------------------------

5. 35, 45, 1, 31, 19 ------------------------------

6. 18, 9, 69, 47, 568 ------------------------------

7. 51, 17, 25, 32, 4 ------------------------------

8. 23, 14, 54, 4, 55 ------------------------------

9. 11, 22, 33, 44, 57 ------------------------------

10. 15, 85, 644, 47, 44 ------------------------------

Exercise 3 :

ARRANGE THESE NUMBERS IN ORDER FROM ASCENDING TO DESCENDING FORM :

1. 5, 45, 478, 4, 5 -------------------------------

2. 12, 10, 6, 44, 68 -------------------------------

3. 45, 66, 4, 3, 100 -------------------------------

4. 555, 2, 22, 66, 20 -------------------------------

5. 45, 56, 22, 99, 4 -------------------------------

6. 12, 51, 58, 77, 1, 5, 4 -------------------------------

7. 51, 17, 25, 32, 555 -------------------------------

8. 23, 14, 54, 4, 55, 4 -------------------------------

9. 11, 22, 33, 44, 57, 2 -------------------------------

10. 15, 85, 644, 47, 584 -------------------------------

MODULE 3

Exercise 4 :

FILL IN THE BOXES

1. 96 = ☐ + 32
2. 14 = ☐ + 9
3. ☐ = 15 + 4
4. 18 = ☐ + 9
5. 9 = 3 + ☐
6. 55 = ☐ + 40
7. ☐ = 10 + 30
8. ☐ = 19 + 5
9. 56 = ☐ + 20
10. ☐ = 40 + 45

Example 2 :

5+5+5+5+5+5 = 30 = **10 X 3** = 30

2+2+2 = 6 = **3X2** = 6

Exercise 5 :

8+8+8+8 = = 8 X 4 =

7+7+7 = = = 27

14+14+14+14 = = =

6+6+6+6+6 = = =

3+3+3+3+3 = = 10 X 3 =

12+12+12+12 = = =

= = 2 X 4 =

☐ = 6 = ☐ = 6

☐ = ☐ = 7 X 7 = 49

9+9+9+9 = ☐ = ☐ = 36

MODULE 4

Example 3 :

One quarter and one half of the shapes have been shaded:

1/2   1/4

My exercise : Shade the following:

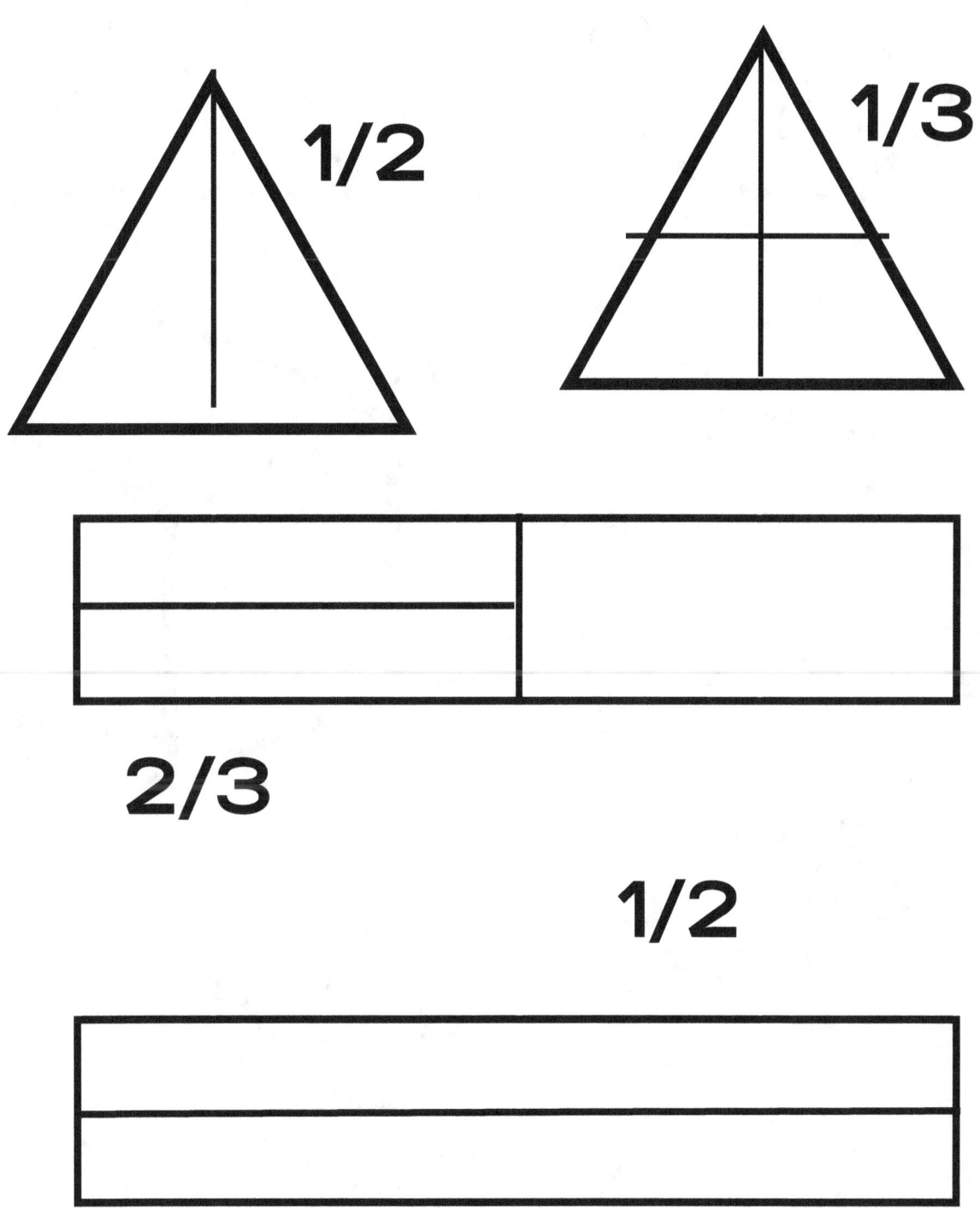

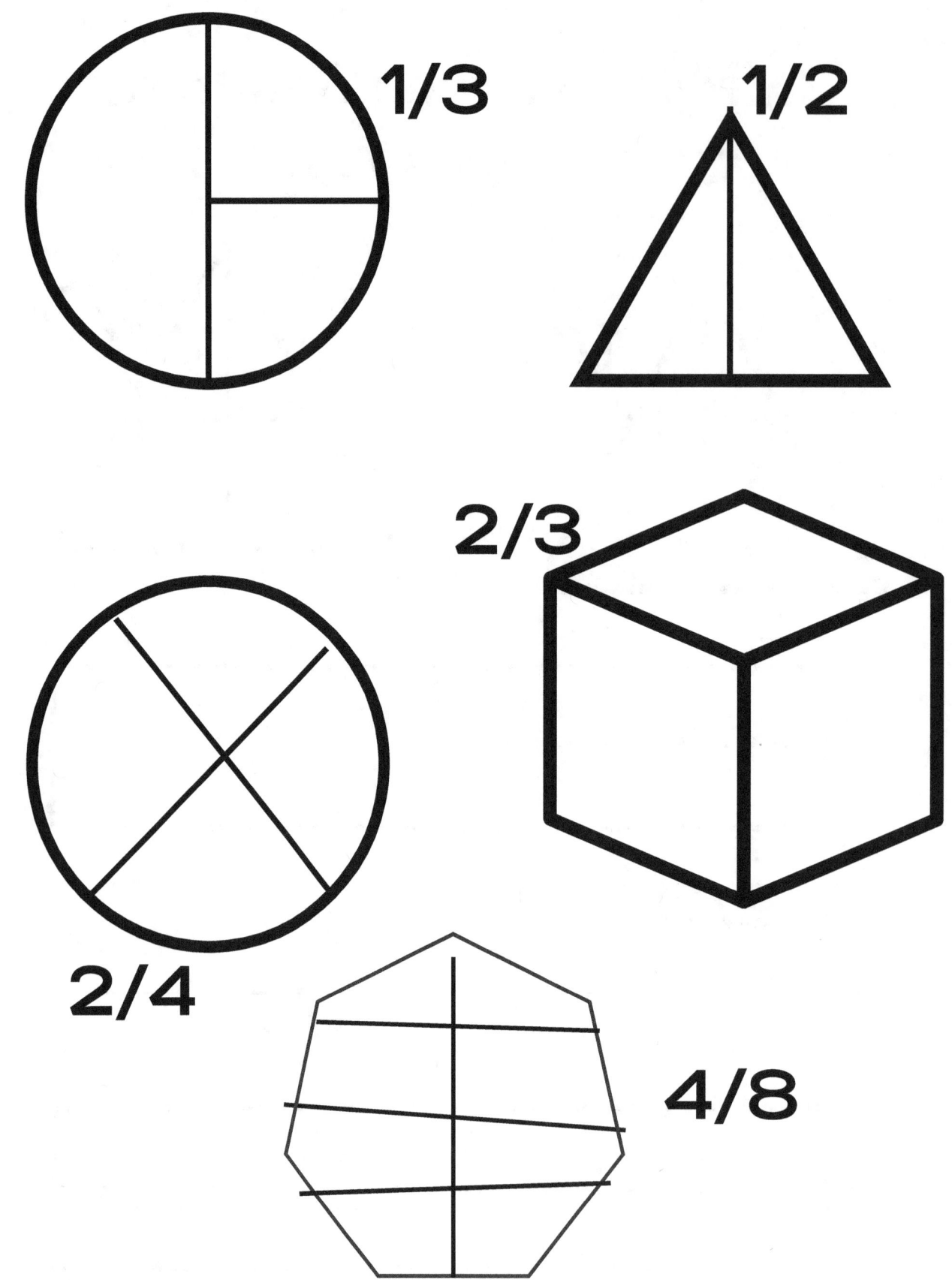

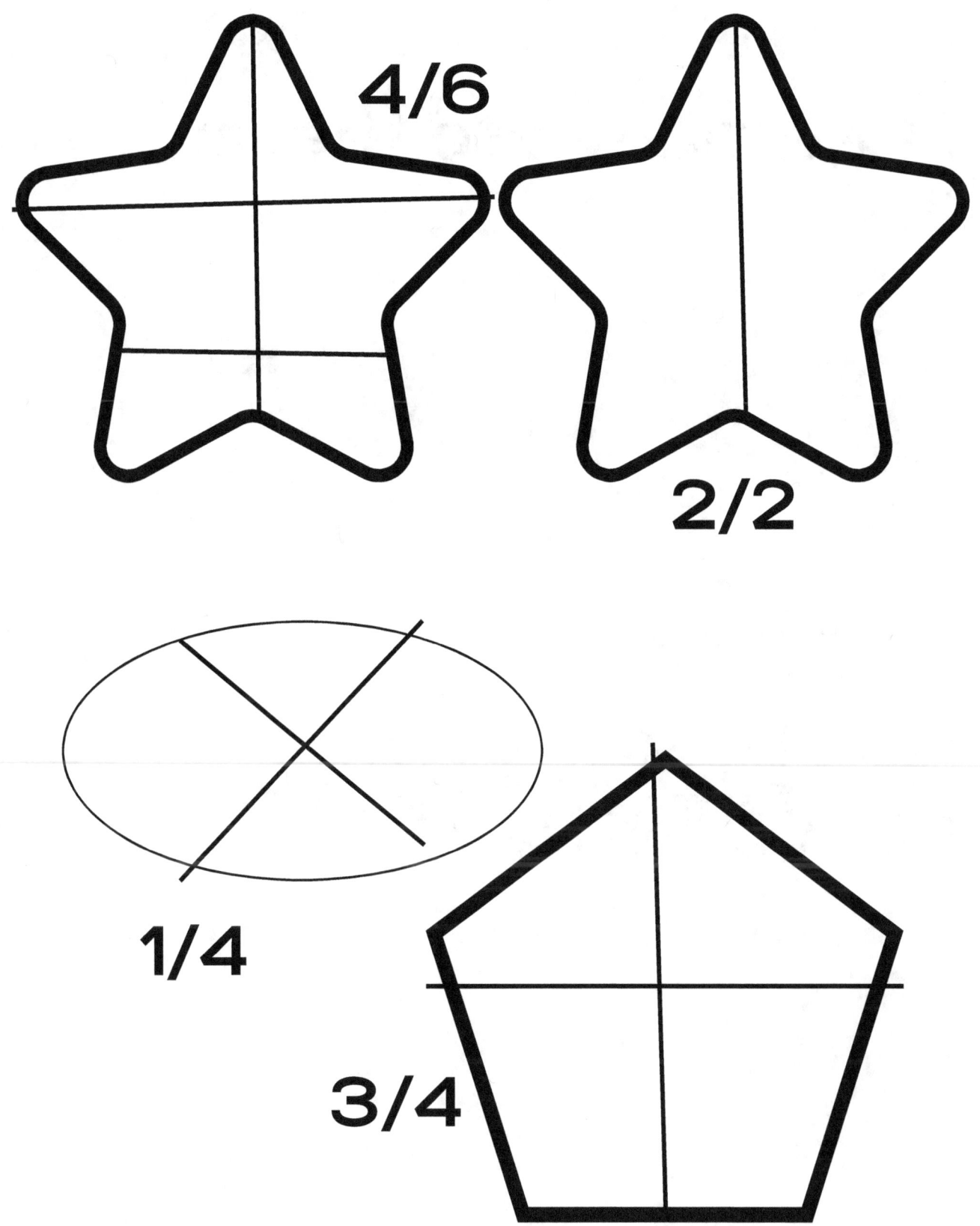

QUIZ TIME

How many are there:

MODULE 5

Work out the following;

$\dfrac{3}{3} \times 24 =$

$\dfrac{4}{5} \times 20 =$

$\dfrac{90}{3} \times 9 =$

$$\frac{14}{7} \times 7 =$$

$$\frac{40}{10} \times 2 =$$

$$\frac{24}{100} \times 8 =$$

Example 4 :

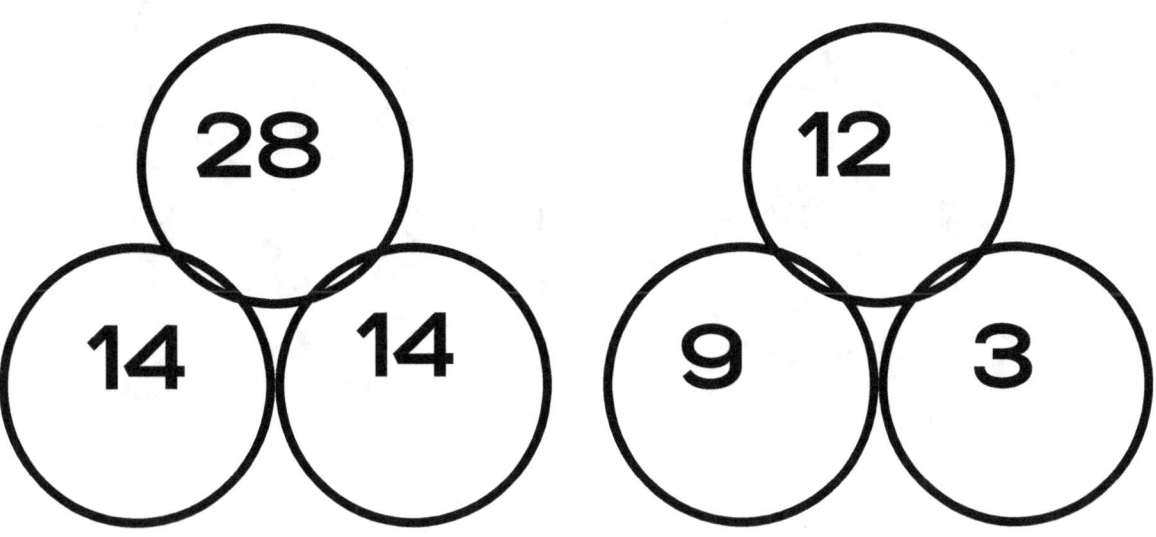

Exercise 6 :

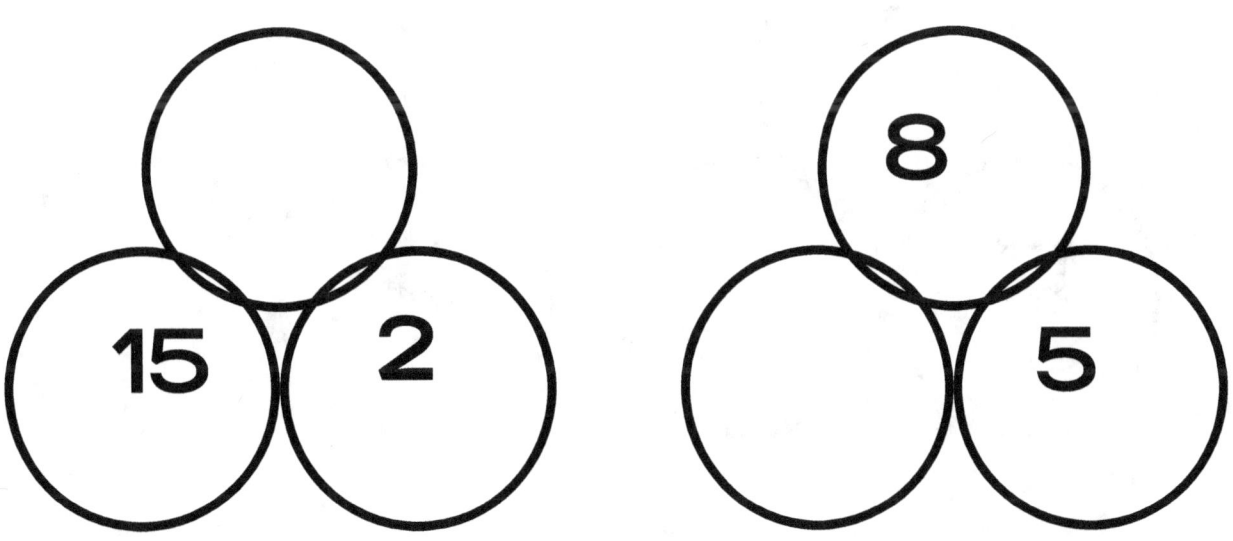

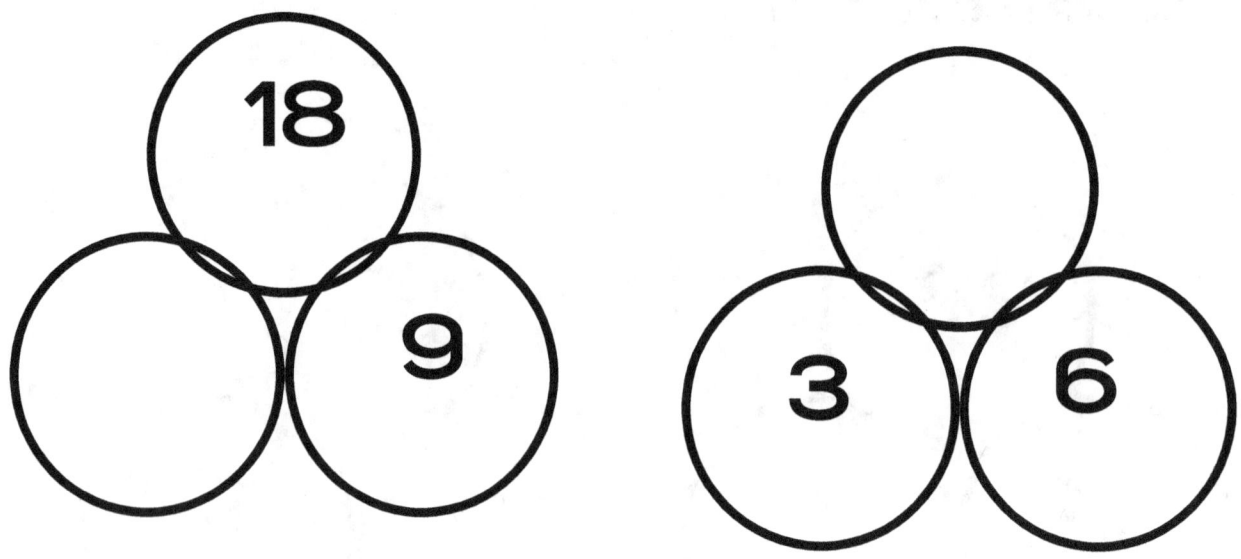

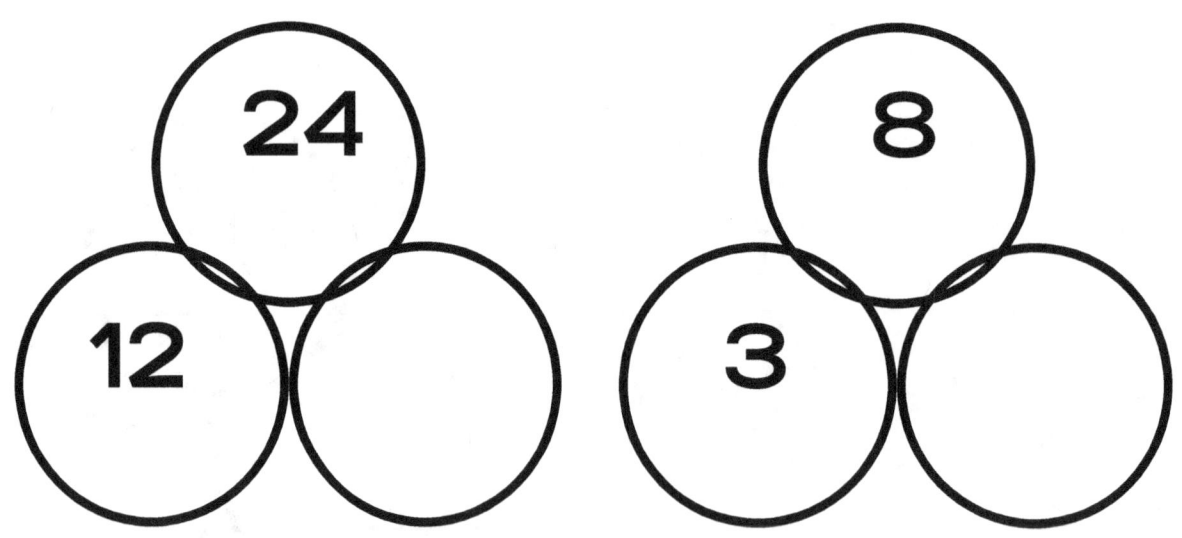

Worksheets

Worksheets

Worksheets

Worksheets

Worksheets

Worksheets

www.ingramcontent.com/pod-product-compliance
Lightning Source LLC
Chambersburg PA
CBHW081529240526

45465CB00029B/2812